Open Gold Claim Finding & Filing Guide©

By
Captain Gene

http://www.opengoldclaims.com

Warning: This Document is Copy Righted

Open Gold Claim Finding & Filing Guide

Table of Contents

Introduction

This guide is the combination of over 5 years worth of research and 10 years of experience. I guarantee you cannot find all this information compiled anywhere else into a single manuscript.

Included in this guide is a step by step process to follow, a check list to keep you on track, links to vital websites and a wealth of information, tips and advice on how, without help from anyone, you can find, stake and file on your own gold claim.

Your claim may or may not house the pot of gold at the end of the rainbow, but it certainly does give you more control over the pot if you do happen to find it.

You and only you are in control – no one can shut you down because your time is up. More importantly, no one else can haul your gold away without your permission. In other words, it's all in the bank in *your* name - the creek bank of course.

If you never gain another thing in this world, if you don't have a high paying job, if you don't have a great retirement plan, whether the economy tanks or soars as long as you fulfill your responsibilities, having your own gold claim means you have invested in a tangible piece of potential wealth.

Open Gold Claim Finding & Filing Guide

Unlike the stock market, you don't have to depend on an investment that can disappear in a mili second. Your right to these minerals is "rock solid".

Your claim is highly valuable and secure for several reasons.

- First, a claim is legally considered to be "real property". Therefore, unless you decide otherwise, your gold claim will always belong to your estate
- Your claim may be leased or rented to others, or used for bartering for goods or services
- You may let other people work your claim for a percentage of the gold recovered. In fact, some miners let other people do their prospecting for them using that very technique.
- You also have the right to sell your claim if you so desire

- Virtually every nation in the world recognizes gold as a commodity, so your claim has global market value
- With a sufficient assay your claim may also have loan value

Now, let's get busy and find, stake and file on you a claim.

Chapter One

Where Am I Going?

The absolute first thing you need to decide is where you want to look for your future gold claim. You may be surprised to learn that not only can gold be found on public lands – such as state and federal land (including National Forest Service Lands), but there is actually gold on private lands as well. By private I mean other people's claims.

Under certain circumstances, claim owners may be willing to rent, lease or sell you mining rights to their claims. Such arrangements may last only a day (pay to play operations), or full mining seasons leased by you from a state claim owner with an entrepreneur spirit, or even the purchase of a patented gold claim.

Please let me give you a word of caution if you plan to buy a claim. First, there is little need to buy a claim, in this writers opinion, when there are so many open gold claims available just waiting for you to find them. Second, if you are buying a gold claim make sure you verify the claim's ownership as well as its true value.

The scope of this guide will be to focus on assisting you in finding and filing on your own claim. The decision at hand is quite simple. Will you choose to climb over the mountain

Of governmental red tape and other obstacles placed strategically in your way, and do the hours and hours of
 research yourself; or will you choose to simply buy my list of thousands of open gold claims just waiting for you to go straight to them, stake them, file the paper work, and begin your prospecting and mining operation immediately?

To nudge you in the right direction I would like to bring a couple of important matters to your attention. First, the price of gold is higher than it has ever been. High gold prices make getting your own gold claim and putting it to work quickly an essential part of the economics of owning a gold claim. Secondly, although I have compiled a list of thousands of gold claims ready and waiting for you to choose from, I also have an ever building list of miners signing up to take them. Put simply, the "rush is on".

Which ever choice you make, this guide is designed to point you in the right direction and give you a check list to help you find your way.

Once you have decided the general location you want to have your gold claim in you will need to find out if there are any open lands on which you can prospect and stake your claim. So, the first step is to make your list of the areas you want to explore for claim possibilities.   The following chapters give you step by step measures to take to narrow your list and hunt down the exact location of your claim beginning with state lands.

Chapter Two

State Claims

Not all states with land available for public mining will be included in the scope of this booklet. However, for your convenience we have included basic information pertaining to 42 different states which may allow for prospecting and filing gold claims on public lands within their jurisdiction.

Below you will find the name of the state and a hyperlink to various state offices managing their mining interests.

- Alabama: Geological Survey of Alabama is the mining claims contact for the state of Alabama   http://www.gsa.state.al.us

- Alaska: Department of Natural Resources in the mining division. http://www.dnr.state.ak.us/mlw

- Arizona: This state's governor has closed down the department that attends to mining in the State of Arizona. We will update this matter on our website as further developments occur. In the meantime, this is the only contact we have http://www.azgs.az.gov

- Illinois: Mining lands within the State of Illinois is managed by the Illinois State Geological Survey office and may be contacted at http://www.isgs.uius.edu

- Indiana: The Indiana Geological Survey office manages mining claims for the State of Indiana and their contact is http://www.igs.indiana.edu

- Iowa: The Iowa Geological Survey office manages the mining interests for the state of Iowa and they may be contacted at http://www.igsb.uiowa.edu

- Kansas: Mining claims inquiries are managed by the Kansas Geological Survey office and you may contact them at http://www.kgs.ku.edu

- Kentucky: The Kentucky Geological Survey office manages mining lands in Kentucky and may contacted at their website http://www.uky.edu/kgs

- Louisiana: Contact the Louisiana Geological Survey office for mining lands at http://www.lgs.lsu.edu

- Maine: Mining concerns for the State of Maine are managed by the Maine Geological Survey office at http://www.maine.gov/dot/nrimc/mgs/mgs.htm

- Maryland: The Maryland Geological Survey office manages mining lands in Maryland and they may be contacted at http://www.mgs.md.gov

- Michigan: Contact the Michigan Department of Geology for mining lands inquiries at the following website http://www.michigan.gov/dnr

- Minnesota: Mining lands are managed for the State of Minnesota by their Department of Natural Resources and they may be contacted at http://www.state.mn.us

- Montana: The Montana Bureau of Mines manages mining lands in the State of Montana and may be contacted at http://www.mbmg.mtech.edu

- Nevada: The two mining management offices for the State of Nevada and their contact information is: Nevada Bureau of Mines http://www.nbmg.unr.edu
  Nevada Minerals Division http://www.minerals.state.nv.us

- New Jersey: The New Jersey Geological Survey office manages the mining lands in the State of New Jersey contact them at http://www.state.nj.us/dep/njgs
- New Mexico: The Mining and Minerals Division in the State of New Mexico manages mining claims contact them at http://www.emnrd.state.nm.us/mmd

- New York: The New York Geological Survey office responds to mining lands inquiries from this website http://www.nysl.nysed.gov

- North Carolina: Mining claims inquiries are managed by the North Carolina Geological Survey office and their website is http://www.geology.enr.state.nc.us

- North Dakota: The North Dakota Geological Survey office manages public mining claims inquiries and their web address is http://www.state.nd.us/ndgs

- Oklahoma: You may contact the Oklahoma Geological Survey office for mining information in the State of Oklahoma at http://www.ogs.ou.edu

- Oregon: The Oregon Geology and Minerals office manages mining operations within the State of Oregon contact website http://www.oregon.gov/dogami

- Pennsylvania: The Pennsylvania Conservation and Natural Resources office manages mining lands at this website http://www.dcnr.state.pa.us/topogeo

- South Carolina: The South Carolina Natural Resources office manages public mining applications at this location http://www.dnr.sc.gov/geology

- South Dakota: The South Dakota Natural Resources office is responsible for managing mining lands in South Dakota contact here http://www.sdgs.usd.edu

- Tennessee: For mining claims inquiries in Tennessee contact the Tennessee Division of Geology at this website http://www.tennessee.gov/environment/tdg

- Texas: The Texas Bureau of Economic Geology manages public mining lands inquiries from this website http://www.beg.utexas.edu

- Utah: Utah public mining lands are managed by the Utah Division of Oil, Gas and Mining office and may be contacted here http://www.geology.utah.gov

- Vermont: The Vermont Geological Survey office manages mining claims from the following website http://www.anr.state.vt.us/dec/geo/vgs.htm

- Virginia: Public mining claims are processed by the Virginia Department of Mines office and they may be contacted at http://www.geology.state.va.us

- Washington: The State of Washington's mining operations are managed by the Washington Division of Geology and Earth Resources and may be contacted at their website http://www.dnr.wa.gov/geology

- West Virginia: The West Virginia Geological and Economic Survey office manages the public mining lands from this website http://www.wvgs.wvnet.edu

- Wisconsin: The State of Wisconsin's mineral mining lands are protected by the Wisconsin Geological and Natural History Survey office and you may contact them at this website http://www.uwex.edu/wgnhs

- Wyoming: Public mining claims applications are available through the Wyoming Geological Survey office at this website http://www.wsgs.uwyo.edu

Note: Thanks to the Fourteenth Amendment of the U.S. Constitution, travel and the right to own property and pursue happiness among the 50 United States of America – and its territories, is guaranteed; therefore, one does not have to be a resident of any particular state (that allows for prospecting, staking and filing a gold claim) in order to participate in the mining process of that state.

Chapter Three

Federal Claims

The General Mining Act of 1872 gave all adult citizens of the United States the right own and operate a mining claim on designated federal public lands.

There are two types of mineral mining claims available on federal land and they are, lode mining claims (those found in hard rock) and placer mining claims (minerals found in gravel or sand).

Minerals that may be mined on public, federal lands are those such as silver, gold, platinum, zinc, lead and copper.

The federal government allows people to stake a mining claim whether or not they have made a mineral discovery; however, it is time sensitive, so don't waste your time waiting for an earthquake to shake the gold out of the ground.

A very important point for now is that claim applicants understand that there are many differences between state and federal mining rules and regulations, fees and filing processes. For example, on most state owned mining claims miners can work with a suction dredge that has an intake nozzle of up to 6 inches whereas, on federal mining claims the limit is only 4 inches or less. Please realize that on some streams that have fish habitat dredging may be highly regulated or may even be forbidden. Violators are heavily fined, so be advised.

Additionally, on federal lands the annual maintenance fees are at this publishing, $140 with no royalty fees being assessed from the miner. Conversely, most states have a lower annual maintenance fee, but require royalty fees from miners for the minerals recovered.

The Federal Bureau of Land Management (BLM) is the primary governing entity that manages mining on federal lands. That certainly does not preclude the many other federal agencies that may have a permitting interest in the mining process such as the Environmental Protection Agency (EPA) or the U.S, Core of Engineers for example.

The days of miners simply taking their pack mule, pick and shovel and heading to the hills have long ago disappeared, so it is my duty to inform my readers that there is more to successfully mining a gold claim than simply finding, staking and filing on it.

For my readers' convenience I have placed a non exhaustive check list at the end of this guide that includes some of the state and federal agencies a modern day miner may find him/herself face to face with.

Chapter Four

Data Mining

In this writer's opinion, one of the greatest challenges most miners face is that of data mining.

Finding a claim, whether state or federal, can be a daunting task for several reasons. Confusing red tape is what makes buying an already established gold claim so appealing.

Many people would just rather pay someone thousands of dollars for an established mining claim. People believe it's easier. The claim is located, filed upon and properly staked.

One of the main functions of this guide is to help miners get beyond this mentality and locate their own claim. We know how frustrating it can be to search and search for hours, weeks and even years for a gold claim. Like many others, we did not have the money to go out and buy a ready made claim.

Although we sell a list that has thousands of FREE claims on it that will have you out the door in a matter of minutes, a person with just a little bit of tenacity can mine this mountain of maps, case files, status plats, surveys, land records and the many other research tools available to them whereby they can locate their own prospective gold claim.

As mentioned before, the first step is to decide where you would like to begin your search for a gold claim

The scope of this guide does not include an in depth tutorial on how to use the BLM's Geo-Locator or you state's online mapping tool; however, if you will go to the following site there is extensive helps and tips sections that will explain in plain English how these mapping resources work. http://www.geocommunicator.gov

My job is to point you in the right direction. There are many helpful people who work at these mining management offices, so if you do get lost don't be bashful about asking for help.

The Data Mining Process:
- Decide where you want to search
- Get a paper map of that exact area (an atlas is not usually the best tool for local searches)
- Are you looking for a state claim or federal claim? Find state / federal public lands that are open for mining by going to mining and mineral websites as listed throughout this guide.
- Choose claim type. Prospecting? Traditional? Other.
- How is the claim types sectioned off - by acreage, feet, legal description, or aliquots (see terms)?
- In other words, if the state you're searching in uses MTRS (see terms) to record its mining claims, then ***getting a map with those attributes to narrow your search with is absolutely essential***. You will need to learn to use the available locator systems, as well. If you want to save yourself a big research

headache, ***go to the USGS website and print out a paper map that shows the Township, Range and Section of the property you want to locate your claim on.*** ***Next go to*** http://www.blm.gov/cadastral/meridians/meridians.htm ***to get the Meridian the property is located in.*** In my opinion, this is a must. Armed with this locator information you should be able to locate any piece of property within the United States.

- Use the state's or federal mining claims locator system to start looking for established claims and seeking out an opportunity to possibly find a near by open claim and become a neighbor.

- If you know the mining history of the area, the legal description or claim identifiers of the parcel you want to prospect, you can usually take that information to the state or federal locator system to obtain the GPS coordinates to the property. Those coordinates will inevitably be used to stake and post your gold claim.

- If you are data mining for a federal claim your job is probably going to be a lot easier. You can use the BLM's GeoCommunicator website to find public lands that have an excellent potential for mineral recovery. http://www.geocommunicator.gov

- An excellent resource for frequently asked questions about filing a mining claim on federal lands is at this website http://www.blm.gov/wo/st/en/prog/more/non-energy_minerals/solid_minerals_brochure.html

- Keep in mind that although many individual states have their own mining programs, there are federally administered lands in only 19 States where you may locate a mining claim. States having available land for staking a claim, in alphabetical order are: Alaska, Arizona, Arkansas, California, Colorado, Florida, Idaho, Louisiana, Mississippi, Montana, Nebraska, Nevada, New Mexico, North Dakota, Oregon, South Dakota, Utah, Washington, and Wyoming.

- The BLM'S Legacy Rehost 2000 system is a searchable database of public records con-cerning BLM's land and mineral use. Excluding Alaska and Hawaii, these records include unpatented mining claims, activities concerning permits, leases, licenses, conveyances, and other authorizations. Search the LR2000 website under reports section here **http://www.blm.gov/lr2000** .

- If you find this process to be too intense for one reason or the other, then go to our website where we offer a list that has thousands of open gold claims waiting for you to pick as many as you want, stake them, file the paperwork, and start mining. **http://www.opengoldclaims.com**

Chapter Five

**The Big Trip
To Stake or Not to Stake?**

Once you have selected a prospective claim site the next step is to physically go out to the property. Different states have different rules about prospecting and staking your new gold claim, so by now you should be armed with the _Fact Sheets_ that describe how to properly stake your claim in that particular jurisdiction.

For the purpose of this guide I will give two examples of how to stake a gold claim.

STAKING A MINING CLAIM OR SITE

Federal law simply specifies that claim boundaries must be distinctly and clearly marked to be readily identifiable on the ground. The mining laws allow states to establish their own laws regarding the manner in which mining claims and sites are located. The miner must comply with both federal and state requirements. THIS is extremely important folks. Be sure you comply with *BOTH* state and federal staking requirements. Don't just think you can comply with federal only because your claim is located on federal lands. If your claim is ever challenged, its being properly staked will be monumental (pardon the pun) to the successful defense of the claim. Most states do have statutes and regulations that add to the General Mining Law of 1872, as amended, concerning the actual staking and recording of mining claims. These staking requirements include the placement, size, and acceptable materials for a corner post or discovery monument.

Because it is cost prohibitive to include the rules of all the states that have mining claim availability, I will pick the state I live in to use as a sample guideline for claim staking requirements.

How to properly stake a mining claim in the State of Alaska

Open Gold Claim Finding & Filing Guide

FILING A TRADITIONAL CLAIM

- Use 4 posts at least 2 ½ inches in diameter and 3 foot tall made of wood, metal, PVC piping, a blazed tree or a rock cairn
- Place the posts on the NE,SE, NW, SW corners of your claim no further than 1320 ft. apart
- Post your notice, Certificate of Location, on the NE corner post with the following information on it:
  1. Claim name/number
  2. Date of posting
  3. Length and Width of claim in feet
  4. Drawing of claim in relationship to surrounding properties
  5. Name, Address and telephone number of claim owner

MTRSC or PROSPECTING CLAIM

- Use above named posting material information
- Place posts at the aliquot (see terms) corner locations for a ¼ section location (for prospecting sites) or either a ¼ / ¼ or ¼ section location for MTRSC claims. The aliquot sections are based on GPS or topographical map quadrants within a section.
- The same marking notice list given above for the traditional claim, also applies to the MTRSC and PROSPECTING claim postings

Further, the State of Alaska, unlike many other states, requires that all four corner posts be marked with the name of the location as well as being numbered sequentially beginning with the NE corner post proceeding to the SE and then the SW and finally the NW.

If you really want to stave off claim jumpers here are some suggestions pertaining to staking your claim.

1. Take pictures of the NE post and include in the picture
   - An identifying landmark on the property
   - Include yourself, your family, your dog etc in the pictures – the more witnesses the merrier
   - The front page of a recent Newspaper can help you prove the date of posting

Additional to taking pictures of the actual staking of your claim, *BE SURE* you use the correct form and fill it out properly. Throughout the years I've seen some very illegible Certificate of Location forms with some pretty vague descriptions drawn/written on them. Incomplete documents require a lot of effort to defend if a dispute or challenge ever arises. I can't stress enough the importance of properly MARKING your claim.

Chapter Six

Filing & Recording Your Documents

Federal Claims

Once you have completed your research to locate your gold claim and properly staked your claim, then you are ready to file your claim form Certificate of Location.

For a federal claim you have 90 days to make such a filing; however, this writer strongly urges you to move toward that recording office as fast as the law allows. Hastily filing on your location is another safeguard against claim jumping.

All location notices must be recorded with the local state recorders office in the district the claim is located.

Additionally, a copy of that recordation should be forwarded to the BLM to be included in the mining claim case file as well.

State Claims

For a state claim the time frame for filing may vary, but again in our example for the State of Alaska, a miner has only 45 days after posting their location to record their Certificate of Location in the appropriate recording district.
The Division of Mining, Land and Water provides a form

for such recording; therefore, prior to taking/sending your notice to the recorders office it is imperative that you get the claim approved by the Department of Natural Resources Public Information Center at which time you will pay your rental fees.

Establishing this paper trail is an absolutely necessary part of your staking a claim of your own. Again, just like the federal process, this writer strongly urges you to make haste getting all paper work filed to prevent possible confusion in the event another person has ideations about the same claim location.

It is essential that a claim locator/miner become familiar with the rules and regulations pertaining to their staking, filing, mining and maintaining their claim.

You may or may not realize that other people (in some states) can cross file on top of you for the same claim. People rudely realize they have lost their claim because they forgot to file/pay their annual fees. In the event that you fail to file your annual maintenance forms, and your claim defaults the claim can automatically go to the next claimant.

It may appear to an outsider that mining is simply grabbing a gold pan, a pick and a shovel and heading into the mountains, streams or river beds seek their fortune.

That train of thought could not be farther from the truth. Mining in the United States, even recreational mining, has become highly regulated with more laws being formulated annually. I hope you'll be diligent about protecting your rights to file your own gold claim at a reasonable cost.

All the Best,

Captain Gene

CHECK OUT MY SPECIAL OFFER at THE END of THIS GUIDE!

Terms You May Need

Aliquot Part – Exact proper divisor such as ¼ section (160 acres of land) or ¼ / ¼ (40 acres of land)

Certificate of Location – Official document usually duplicated. One copy may be placed at the claim's NE corner and the other copy is filed with the recorder's office.

Fees – Various local, state and federal fees that must be paid in order to obtain permits, file gold claims or maintain your claims

GPS – Global Positioning System using Longitude/Latitude coordinates (Be sure your GPS is set up to match the mapping system you are using. Otherwise, you may find yourself walking around and around in circles out in the field. You may need to change the NAD in your GPS. Contact the help line of the manufacturer if need be.)

Legal Description called MTRSC – Meridian, Township, Range, Section and Claim

Mining Locations – 3 basic types of gold mining locations are: Prospecting Sites, Leasehold Locations and Mining Claims

Patented Claims – Claims that have had the property – not just the mineral rights - conveyed into the ownership of the claim owner. It allows the claim owners to treat the land as
their own. It's the same as if they purchased the property.

Recording District – District Recording office that your claim Certificate of Location must be physically filed in

Suction Dredging – Mechanized equipment with a regulated intake nozzle and engine. State claims may allow up to a 6inch nozzle for recreational mining, however the federal government only allows up to a 4 inch nozzle for recreational gold mining http://www.opengoldclaims.com

Permitting/Governing Agencies

The following is by no means meant to be an exhaustive list of governmental agencies a miner may encounter on a mining claim application. Therefore, please consider it simply a list compiled for your convenience to which your location may add or subtract.

1. Bureau of Land Management
2. Department of Natural Resources
3. U.S. Fish and Wildlife
4. State Fish and Game
5. Environmental Protection Agency
6. U. S. Army Core of Engineers
7. State Department of Environmental Conservation
8. U. S. Parks Service
9. U. S. Forest Service
10. State Parks Department
11. State Forest Service
12. State Department of Revenue

Check List

You are encouraged to Print This Page

_____Decide where you want to begin looking for your claim

_____Make a list of Areas that interest you

_____Contact the State Office in Charge of Mining Claims to obtain Fact Sheets and ask questions

_____Use State, Federal, or even USGS topo maps to find the area of Interest

_____Use any tools you can find to obtain the legal description of the parcel your interested in. Local area maps, case files, status plats, surveys, or other land records

_____Contact the Federal Bureau of Land Management (BLM) for potential open claim opportunities

_____Check the BLM's LR2000 system for active claims in the general area of where you are searching for a claim

_____Look for possible open claims near active claims using all available data in the BLM GeoCommunicator

_____Once you locate a potential claim property, it may be possible to record the Longitude/Latitude right off of a State or Federal Mapping System

_____Download Certificate of Location Form

_____Obtain all appropriate local, state or federal permits necessary for recreational mining or prospecting of your potential claim

_____Go to Claim Area to visually inspect the potential claim

_____Prospect/Sample your new claim using equipment such as a metal detector, a gold pan, sluice box or even a shovel to dig a test hole which may assist in determining the potential value of your gold claim.

_____Using Global Positioning Satellite (GPS) coordinates find the NE, NW, SE, SW corners.

_____Stake all four corners with appropriate staking materials

_____Write Appropriate Location Information on your Certificate of Location form and Post as Directed by State or Federal Regulations

_____Take photographs of your claim for proof of your staking efforts. If possible, include a current copy of a daily newspaper and a land mark in the photo to prove the date, place and time you staked and posted your claim.

_____Take a copy of your Certificate of Location to the state or federal mining claims office for approval and registration of your newly posted Certificate of Location.

_____Pay all appropriate fees

_____Take the original document the mining claims office issues to you over to the local, state or federal recorders office for official recording of the mining claim documentation

_____Check out our current **Open Gold Claims List** and other cool tools at this link All claims on our list are FREE for the taking, but we do charge for the list.

http://www.opengoldclaims.com/

SPECIAL OFFER

For anyone who would like to have a short cut to finding their own gold claim using the, **Open Gold Claims List** (and choose from among the thousands of gold claims available on our list) we're offering the following:

"Trade In" your Finding, Staking & Filing Guide for a full $27 credit toward our **Open Gold Claims List**

Do you actually "trade in" your guide and send it back? *No, you do NOT give it back! However, we give you:

- A FULL $27 credit toward your purchase of the **Open Gold Claims List**.

*There is **NO Time Limit** on This Offer*